INTERNATIONAL INTERIOR DESIGN YEARBOOK 2016

2016
国际室内设计年鉴 ③

RESTAURANT
餐馆

本书编委会 编

中国林业出版社
China Forestry Publishing House

图书在版编目（CIP）数据

国际室内设计年鉴. 2016. 餐馆 /《国际室内设计年鉴2016》编委会编. -- 北京：中国林业出版社，2016.4

ISBN 978-7-5038-8448-1

Ⅰ. ①国… Ⅱ. ①国… Ⅲ. ①餐馆－室内装饰设计－世界－2016－年鉴 Ⅳ. ①TU238-54

中国版本图书馆CIP数据核字(2016)第057494号

本书编委会

◎ 编委会成员名单
丛书主编：柳素荣
编写成员：陈向明　陈治强　董世雄　冯振勇　朱统菁
◎ 丛书策划：北京和易空间文化传播有限公司
◎ 特别鸣谢：《室内设计与装修》杂志社
◎ 装帧设计：北京睿宸弘文文化传播有限公司+LOMO红绿

中国林业出版社 · 建筑家居出版分社

责任编辑：王思源　纪　亮

出版：中国林业出版社　（100009 北京西城区德内大街刘海胡同7号）
网址：lycb.forestry.gov.cn
电话：（010）8314 3518
发行：中国林业出版社
印刷：北京利丰雅高长城印刷有限公司
版次：2017年2月第1版
印次：2017年2月第1次
开本：230mm×305mm　1/16
印张：13.5
字数：200千字
定价：220.00元

CONTENTS 目录

004_ 导言
INTRODUCTION

007_ 餐馆
RESTAURANT

008_ 东方花园饭店
ORIENT GARDEN HOTEL

010_ NOBU DUBAI
NOBU DUBAI

014_ 东方火锅工程
ORIENT POT ENGINEERING

020_ MOJO交互式潮流概念餐厅(商业空间)
MOJO INTERACTIVE FASHION CONCEPT RESTAURANT (BUSINESS SPACE)

026_ 南京博得高尔夫酒店
NANJING BODE GOLF HOTEL

032_ GIUSEPPE ARNALDO & SON'S 餐厅
GIUSEPPE ARNALDO & SON'S RESTAURANT

036_ GOLFSEMPACH
GOLFSEMPACH

042_ GIACOMO
GIACOMO

046_ 银杏金阁酒楼
GINGKO BACCHUS RESTAURANT

050_ NOBU FIFTY SEVEN
NOBU FIFTY SEVEN

054_ NOVOTEL TOWER BRIDGE
NOVOTEL TOWER BRIDGE

060_ PEARLS & CAVIAR
PEARLS & CAVIAR

066_ 扬州牛排城（京华店）
YANGZHOU STEAK TOWN(JINGHUA BRANCH)

070_ THE TRAVELERS' LOUNGE
THE TRAVELERS' LOUNGE

074_ 长春市泉家居酒楼
CHANGCHUN HOUSE RESTAURANT

080_ 在GUGGENHEIM博物馆的赖特多万
THE WRIGHT AT THE GUGGENHEIM MUSEUM

084_ 厨房制造
MAKE IN KITCHEN

090_ 豆捞坊古北店
BEAN GUBEI BRANCH

094_ 港鼎汇火锅料理
GANGDINGHUI POT CUISINE

102_ 工体享宴火锅店
GONGTI XIANGYAN POT

110_ 高雄新天地雅悦会馆
KAOHSIUNG NEW EARTH YAYUE CLUB

114_ 黄记煌三汁焖锅中心城店
HUANGJIHUANG THREE JUICE STEW POT CENTRAL RESTAURANT

118_ 鸿霖时尚餐厅
HONGLIN FASHION RESTAURANT

122_ 南京东部山庄抱朴园
NANJING EASTERN VILLA BAOPU PARK

126_ 闽都别景·燕鲍鱼翅会所
MINDU BIEJING · ABALONE AND FIN CLUB

134_ 望湘园金桥店
XIANG GARDEN JINQIAO BRANCH

138_ 星愿时尚餐厅
STAR WISH FASHION RESTAURANT

144_ 扬州牛排城（四望亭店）
YANGZHOU STEAK TOWN(SIWANG PAVILION BRANCH)

148_ 尚品世家
SHANGPIN FAMILY

152_ WGV CAFETERIA
WGV CAFETERIA

154_ 美丽华酒店员工餐厅
THE MIRA STAFF RESTAURANT

158_ 新紫阳大酒楼
NEW ZIYANG RESTAURANT

164_ 隐泉HATSUNE日式料理
SECRET SPRING HATSUNE JAPANESE CUISINE

170_ 扬州牛排城（珍园店）
YANGZHOU STEAK TOWN(ZHENYUAN BRANCH)

174_ 天度会
TIANDU CLUB

178_ 台米台菜 (五四店)
TAIMI TAICAI (FIVE FOUR BRANCH)

182_ 食为天菜馆
FOOD IS GOD RESTAURANT

184_ 食尚廊桥火锅
FOOD BRIDGES POT

190_ LUCK 料理店
LUCK RESTAURANT

192_ 连云港腾轩火锅店
LIANYUN HARBOUR TENGXUAN POT SHOP

196_ 净雅阳光海岸嘉和店
JINGYA SUNLIGHT COAST JIAHE SHOP

200_ 杭州名爵咖啡馆
HANGZHOU MINGJUE COFFEE SHOP

204_ 汉府餐厅
HAN RESTAURANT

208_ 干锅轩
GRIDDLE RESTAURANT

212_ 夜朗蛙工程
NIGHT FROG ENGINEERING

216_ 会仙莜面馆
HUIXIANYOU NOODLES

220_ THE NIGHT MARKET
THE NIGHT MARKET

222_ SECOND HOME KITCHEN AND BAR
SECOND HOME KITCHEN AND BAR

224_ APOSTROPHE
APOSTROPHE

226_ 大渔铁板烧喜荟城店
DAYU IRON XIHUI RESTAURANT

228_ 北京湘君府酒楼
BEIJING XIANGJUN RESTAURANT

236_ YELLOW DELI
YELLOW DELL

238_ JAMIE'S ITALIAN
JAMIE'S ITALIAN

242_ 巴比肯FOODHALL和酒廊
BARBICAN FOODHALL & LOUNGE

246_ NATURE CAFÉ AND NATURE BOUTIQUE
NATURE CAFÉ AND NATURE BOUTIQUE

250_ SUPPERCLUB SINGAPORE
SUPPERCLUB SINGAPORE

254_ MOC MOC
MOC MOC

256_ ESPRESSO IN EXISTING SPACE IJBURG
ESPRESSO IN EXISTING SPACE IJBURG

258_ 福师傅连锁餐厅
FU MASTER CHAIN RESTAURANTS

262_ 唐山凤凰园烤鸭店唐丰路店
TANGSHAN PHOENIX ROASTED DUCK RESTAURANT-FENGLU BRANCH

270_ 南京东佳会馆宴会厅
NANJING DONGJIA HALL BALLROOM

272_ KOSOMO 咖啡店
KOSOMO COFFEE SHOP

274_ 杭州泰晤士咖啡馆
HANGZHOU THAMES COFFEE SHOP

276_ 湘乐汇莘庄仲盛店
XIAONG XINZHUANG BRANCH

280_ SALON DES SALUTS
SALON DES SALUTS

282_ NEVY
NEVY

284_ P*ONG DESSERT BAR
P*ONG DESSERT BAR

286_ BRASSERIE WITTEVEEN
BRASSERIE WITTEVEEN

290_ WALL & WATER RESTAURANT
WALL & WATER RESTAURANT

292_ RESTAURANT TUSEW SWEDEN RAMUNDBERSER
RESTAURANT TUSEW SWEDEN RAMUNDBERSER

294_ FORNERIA SAN PAOLO
FORNERIA SAN PAOLO

296_ CAFÉ 3
CAFÉ 3

298_ CAFFÈ DEI MUSEI
CAFFÈ DEI MUSEI

300_ ADOUR ALAIN DUCASSE AT THE ST.REGIS
ADOUR ALAIN DUCASSE AT THE ST.REGIS

302_ 海得利餐饮酒店
SEA DELI DINING HOTEL

INTRO-
DUCTION

/ RECORD THE EXCELLENCE PUBLISH THE QUINTESSNCE

/ 记录精英 传播经典

张先慧 /Zhang Xianhui

中国麦迪逊文化传播机构董事长
中国（广州、上海、北京）广告书店董事长
《麦迪逊丛书》主编
Chairman of China Madison Culture
Communication Institutions
President of China(Guangzhou,Shanghai,Beijing)
Advertising Bookshop
Chief Editor of "The Madison Series"

人的一生，绝大部分时间是在室内度过的。因此，人们设计创造的室内环境，必然会直接关系到人们室内生活、生产活动的质量，关系到人们的安全、健康、效率、舒适，等等。随着人们生活水平和审美能力的不断提高，人们更加注重生活环境的设计，对于室内设计的要求更加严格，需求也日益多样化、个性化。这就要求设计师一定要牢牢把握住时代的脉搏和潮流，以独特的眼光，运用与众不同的角度和表现手法进行创意性的设计，以满足人们对室内设计的需求。

然而，一件好的设计作品，不仅与设计师的专业素质和文化艺术素养等联系在一起，更离不开对他人成功经验的借鉴，为此，《国际室内设计年鉴2011》应运而生。

本年鉴秉持以中国大陆、中国香港、中国台湾为主，兼容其他国家与地区参与的原则，主张以创新与发展作为室内设计创作的主旋律，以科学与艺术相结合的审美眼光审视室内设计作品，力求打造全球最具影响力的室内设计行业年鉴，并使其成为各国设计师可以借鉴的经典书籍。

本年鉴征稿消息发出后，世界各地的设计机构与设计师都踊跃参与，大量投稿，投稿数量之多完全出乎我们的意料，最终本年鉴以一套五册的形式面世。

我们用年鉴的形式把当代最具价值的室内设计作品记录下来，传播开去，意在对室内设计文化予以保存的同时，也为读者提供了解当代设计状况及思想交流的平台。

"记录精英，传播经典"，这是《麦迪逊丛书》的宗旨。

希望业界朋友继续关注与支持我们。

One's lifetime mostly passes through in the interior. Therefore, the interior environment will directly involve quality of people's interior life, activities, people's safety, health, efficiency, comfort and so on. Along with the continuous improvement of people's living standard and aesthetic capacity, people pay more attention to living environment design, and their requirement for interior design is more strict, increasingly diverse and personalized. This requires that the designer firmly grasp the pulse of the times and trends, with the special insight, to use the different angles and methods of performance for creative design, in order to meet the needs of people's interior design requirement.

A masterpiece requires not only the link of the designer's professional quality and cultural art accomplishment, but also others' successful experiences. For this reason, "International Interior Design Yearbook 2010" is born at this right moment.

This yearbook gives priority to China Mainland, China Hong Kong and China Taiwan and pays much attention to other countries and areas, and it upholds the spirit that innovation and development should be the theme of interior design and that interior design works should be evaluated in a scientific and artistic perspective. Aiming at becoming the most influential global yearbook of interior design, this book is a classical one in the eyes of designers all over the world.

After the announcement of draft-collecting was spread, we have received so many contributions from the designers and organizations of almost every country. The number was so surprising. Finally, the yearbook is published in a set of five books.

We present the most valuable contemporary interior designs through publishing this yearbook in order to preserve the interior designing culture and provide a platform for readers to know about contemporary designing improvements and to communicate with each other.

"Record Excellence Works, Spread Classical Works" is the tenet of "Madison Series".

It will be our privilege to have your appreciation and support.

RESTA-
URANT
餐馆

餐馆·RESTAURANT

东方花园饭店
ORIENT GARDEN HOTEL

项目资料：
设计单位：北京丽贝亚建筑装饰工程有限公司
设计总监（主创）：曹姗娜
参与设计师：杜春亮 屈勇 卢迪 李昱晓
摄影：徐盟
主要材料：意大利木纹石、古木纹石材、珍珠木、榆木
项目地址：北京市东城区东直门南大街

Project information:
Design Unit: Beijing Libeiya Architectural Decoration Engineering Co.,Ltd.
Design Director(Main Director): Cao Shanna
Involed Designer: Du Chunliang, Qu Yong, Lu Di, Li Yuxiao
Photographer: Xu Meng
Materials: Italian wood stone, ancient stone wood, pearlwood, Elm
Project Address: Beijing East Town District Dongzhi Gate North Street

NOBU DUBAI

项目资料：
设计单位：Rockwell Group
摄影师：Bruce Buck
客户：Alain Ducasse

Project Information:
Design Unit: Rockwell Group
Photographer: Bruce Buck
Client: Alain Ducasse

餐馆 · RESTAURANT

餐馆·RESTAURANT

东方火锅工程
ORIENT POTENGINEERING

项目资料：
设计师：陈旭东 刘秀梅
摄影师：陈旭东

Project Information:
Designer: Chen Xudong, Liu Xiumei
Photographer: Chen Xudong

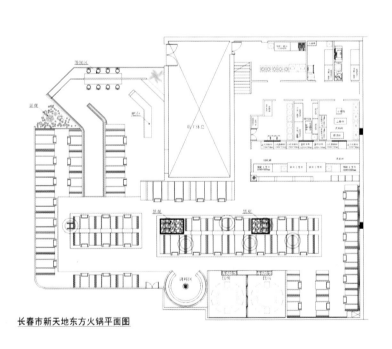

长春市新天地东方火锅平面图

餐馆 · RESTAURANT

餐馆 · RESTAURANT

MOJO交互式潮流概念餐厅（商业空间）

MOJO INTERACTIVE FASHION CONCEPT RESTAURANT (BUSINESS SPACE)

项目资料：
设计单位：木石研室内建筑空间设计有限公司
设计总监（主创）：范赫铄
参与设计团队：木石研室内建筑空间设计有限公司
摄影师：Marc Gerritsen
项目地址：台北市忠孝东路4段1号8F
主要材料：木、玻璃、压克力、电子器材

Project Information:
Design Unit: MOXIE DESIGN
Design Director: Fan Heshuo
Involved Design Team: Mushiyan Interior construction space Design Co., Ltd
Photographer: Marc Gerritsen
Project Address: Floor 8 No. 1 4 Duan Zhongxiao East Road Taipei
Materials: Wood, glass, acrylic, electronic equipment

餐馆 · RESTAURANT

餐馆・RESTAURANT

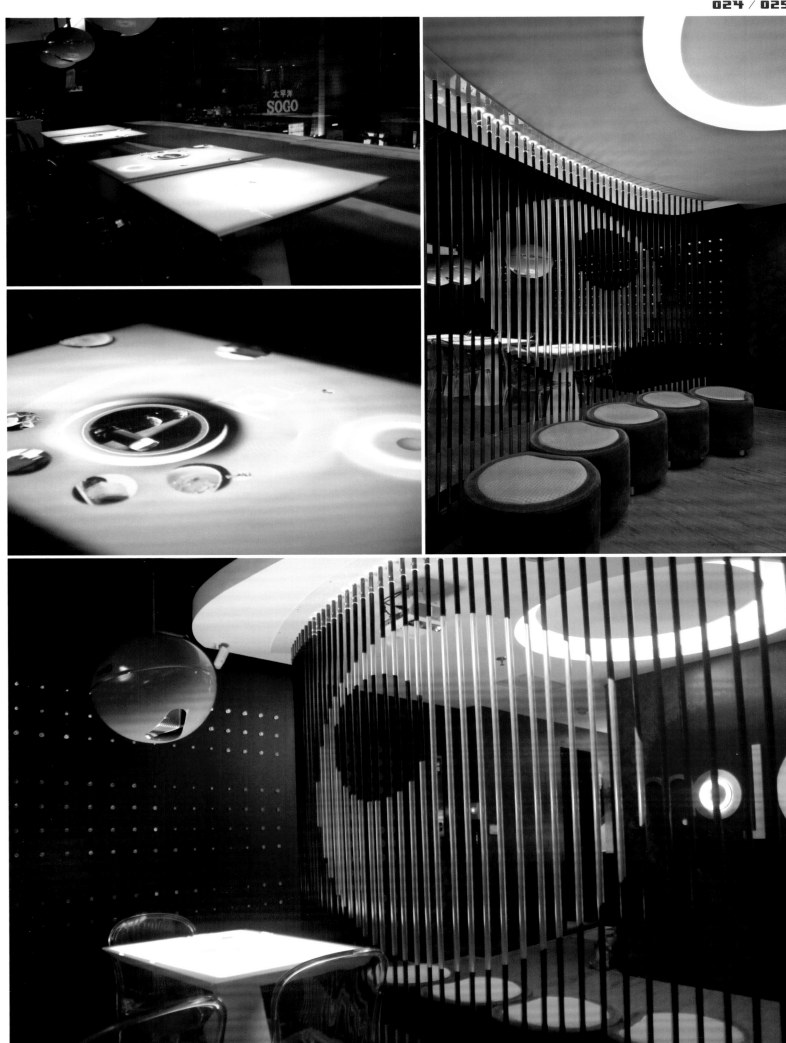

餐馆 · RESTAURANT

南京博得
高尔夫酒店
NANJING BODE GOLF HOTEL

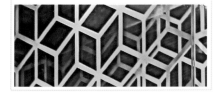

项目资料：
设计单位：南京陆文星装饰设计工作室
设计总监（主创）：陆文星
项目地址：南京仙林
建筑面积：3000m²
主要材料：雕刻玻璃、黑白板是石材、黑檀木、光纤灯、镜面不锈钢等

Project Information：
Design Unit: Nanjing Luwenxing Decoration Design Studio
Design Director(Main Director): Lu Wenxing
Project Address: Nanjing Xianlin
Area: 3000sqm
Materials: Engraving glass, black and white stone, ebony, fiber optic lights, mirror stainless steel.

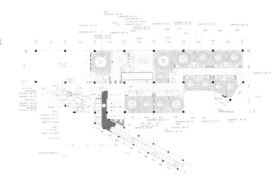

餐馆 · RESTAURANT

GIUSEPPE ARNALDO & SON'S 餐厅
GIUSEPPE ARNALDO&SON'S RESTAURANT

项目资料：
设计单位：Lazzarini Pickering Architects
摄影师：Mattero Piazza

Project Information:
Design Unit: Lazzarini Pickering Architects
Photographer: Mattero Piazza

餐馆·RESTAURANT

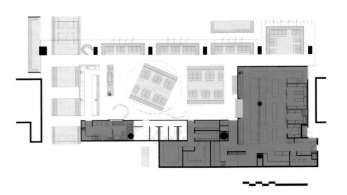

RESTAURANT

GOLFSEMPACH

项目资料：
设计单位：Smolenicky & Partner Architektur GmbH

Project Information:
Design Unit: Smolenicky & Partner Architektur GmbH

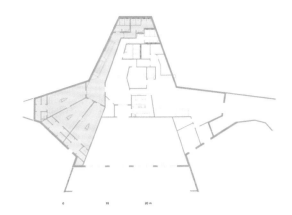

餐馆 · RESTAURANT

餐馆·RESTAURANT

GIACOMO

GIACOMO

项目资料：
设计单位：PLAJER & FRANZ STUDIO
项目管理：mrs. Astrid dressel
摄影师：ken schluchtmann
客户：giacomo natural gmbh
面积：140m²

Project Information:
Design Unit: PLAJER & FRANZ STUDIO
Project manager: Mrs. Astrid dressel
Photographer: ken schluchtmann
Client: giacomo natural gmbh
Area: 140sqm

餐馆 · RESTAURANT

餐馆·RESTAURANT

银杏金阁酒楼
GINGKO BACCHUS RESTAURANT

项目资料：
设计单位：Hotao Chow and Keizo Okamoto
设计师：Patrick Yu（资深设计师）
项目团队：Tang Fei, Li Mei, Florian Niedworok, Tony Wei
ZINTA：Ricky Lauwis, Yu Changong, Li Dehong
摄影师：Golf Tattler: Lai Xuzhu, Oak Taylor Smith
客户：银杏餐厅管理公司
项目地址：成都锦里路2号
完成时间：2008年12月

Project Information:
Design Unit: Hotao Chow and Keizo Okamoto
Designer: Patrick Yu (Senior Associate)
Project team: Tang Fei, Li Mei, Florian Niedworok, Tony Wei
ZINTA: Ricky Lauwis, Yu Changong, Li Dehong
Photographer: Golf Tattler: Lai Xuzhu, Oak Taylor Smith
Client: Gingko Restaurant Management Corporation
Address: No.2 Jin Li Zhong Lu, Chengdu
Completion: December 2008

餐馆·RESTAURANT

餐馆 · RESTAURANT

NOBU FIFTY SEVEN
NOBU FIFTY SEVEN

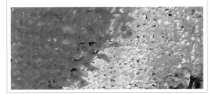

项目资料:
设计单位: Rockwell Group
摄影师: Scott Frances

Project Information:
Design Unit: Rockwell Group
Photographer: Scott Frances

餐馆·RESTAURANT

餐馆 · RESTAURANT

NOVOTEL TOWER BRIDGE
NOVOTEL TOWER BRIDGE

项目资料：
设计单位：Blacksheep
摄影师：Gareth Gardner

Project Information:
Design Unit: Blacksheep
Photographer: Gareth Gardner

餐馆 · RESTAURANT

餐馆 · RESTAURANT

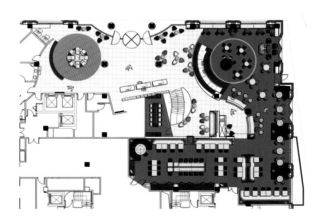

餐馆 · RESTAURANT

PEARLS & CAVIAR

PEARLS & CAVIAR

项目资料：
设计单位：Concrete Architectural Associates
项目团队：rob wagemans, lisa hassanzadeh, sofie ruytenberg, erik van dillen
摄影师：Richard Thorn
客户：Al Jaber
项目地址：Shangri La Hotel Abu Dhabi, UAE
面积：1430m²

Project Information:
Design Unit: Concrete Architectural Associates
Project Team: rob wagemans, lisa hassanzadeh, sofie ruytenberg, erik van dillen
Photographer: Richard Thorn
Client: Al Jaber
Project Address: Shangri La Hotel Abu Dhabi, UAE
Area: 1430 sqm

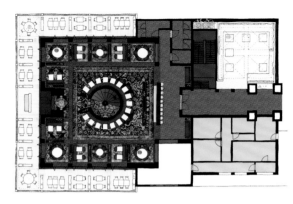

餐馆 · RESTAURANT

餐馆·RESTAURANT

扬州牛排城（京华店）
YANGZHOU STEAK TOWN (JINGHUA BRANCH)

项目资料：
设计单位：上瑞元筑设计制作有限公司
设计师：孙黎明 崔晓曼
摄影师：戴俊峰
面积：250m²
主要材料：橡木夹板、地砖、中花白大理石、铜镜、黑色钢琴漆
项目地址：邗江区京华城路168号
完成时间：2009年5月

Project Information:
Design Unit: Shangrui Yuanzhu Design Making Co., Ltd.
Designers: Su Liming, Cui Xiaoman
Photographer: Dai Junfeng
Area: 250sqm
Materials: Oak plywood, floor tiles, the gray marble, bronze mirrors, piano black
Project Address: Hanjiang Jinghuacheng Road No.168
Completion: May, 2009

餐馆·RESTAURANT

餐馆 · RESTAURANT

THE TRAVELERS' LOUNGE

项目资料:
设计单位: Kinny Chan and Associates
项目团队: KCA Design Team
摄影师: Mr. Kinny Chan

Projects Information:
Design Unit: Kinny Chan and Associates
Project Team: KCA Design Team
Photographer: Mr. Kinny Chan

项目资料:
设计单位: Kinny Chan and Associates
项目团队: KCA Design Team
摄影师: Mr. Kinny Chan

餐馆 · RESTAURANT

餐馆·RESTAURANT

长春市泉家居酒楼
CHANGCHUN HOUSE RESTAURANT

项目资料：
设计单位：长春市绿廊装饰设计公司
设计总监（主创）：付养国
参与设计师：刘达 司海南 王国庆 张铎
摄影：郭义
项目地址：临河街5062号
主要材料：壁纸、银镜、饰面板、装饰画、透光石

Project Information：
Design Unit: Changchun Lvlang Decoration Design Company
Design Director(Main Director): Fu Yangguo
Involed Designer: Liu Da, Si Hainan, Wang Guoqing, Zhang Duo
Photographer: Guo Yi
Project Address: No. 5062 Linhe Street
Materials: Wallpaper, silver mirror, decorative panels, decorative painting, translucent stone

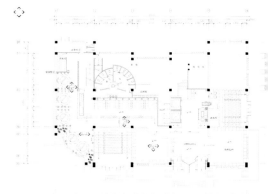

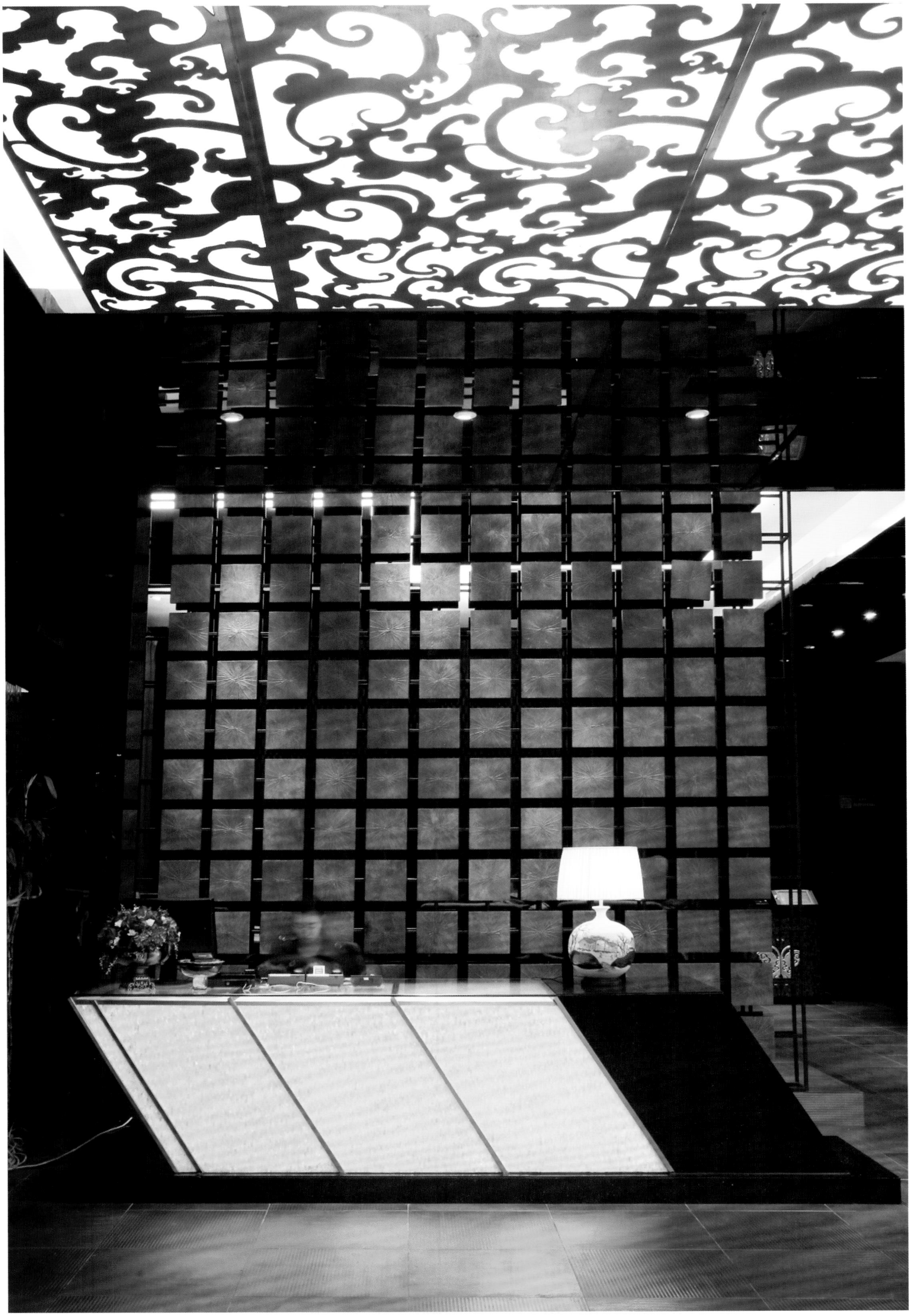

餐馆 · RESTAURANT

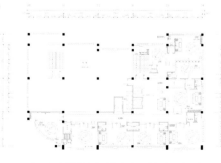
一层平面布置图

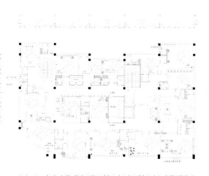
三层平面布置图

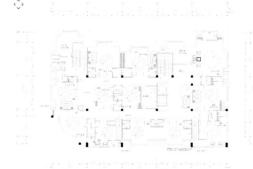
四层平面布置图

餐馆 · RESTAURANT

在 GUGGENHEIM 博物馆的赖特多万
THE WRIGHT AT THE GUGGENHEIM MUSEUM

项目资料：
设计单位：Andre Kikoski Architect
摄影师：Peter Aaron at ESTO

Project Information:
Design Unit: Andre Kikoski Architect
Photography: Peter Aaron at ESTO

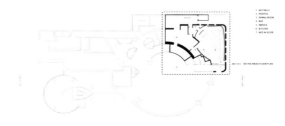

SITE PLAN

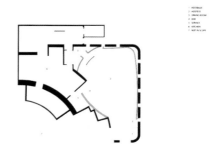

ENLARGED FLOOR PLAN

餐馆·RESTAURANT

厨房制造
MAKE IN KITCHEN

项目资料：
设计单位：PANORAMA泛纳设计事务所
撰稿：潘鸿彬（创办人，泛纳设计事务所）
设计组：潘鸿彬 谢健生 黄卓荣
摄影师：吴潇峰
面积：4000m²
项目地址：中国芜湖
客户：海港饮食策划管理有限公司
完成时间：2010年3月

Project Information:
Design Unit: Panorama Design Office
Word Editor: Fan Hongbin
Design Team: Pan Hongbin, Xie Jiansheng, Huang Zhuorong
Photographer: Wu Xiaofeng
Project Address: China wuhu
Area: 4000sqm
Client: Haigang Diet Planning Management Co., Ltd.
Completion: March, 2010

餐馆 · RESTAURANT

餐馆 · RESTAURANT

餐馆·RESTAURANT

豆捞坊古北店
BEAN GUBEI BRANCH

项目资料：
设计单位：上海沈敏良室内设计有限公司
设计总监（主创）：沈敏良
摄影：蔡峰
项目地址：上海市长宁区古北
主要材料：红镜、紫镜、黄镜、镜面不锈钢镀玫瑰金、墙纸、黑色烤漆钢板线切割
面积：710m²

Project Information:
Design Unit: Shanghai Shenminliang Interior Design Co., Ltd.
Designer: Shen Minliang
Photographer: Caifeng
Project Address: Shanghai Changning District Gubei
Materials: Red mirror, purple mirror, yellow mirror, rose-gold plated stainless steel mirror, wallpaper, paint black steel wire cutting.
Area: 710sqm

餐馆 · RESTAURANT

港鼎汇火锅料理
GANGDINGHUI POT CUISINE

项目资料：
设计单位：南京陆文星装饰设计工作室
设计总监（主创）：陆文星
项目地址：安徽合肥
面积：2000m²
主要材料：黑白板石材、镜面马赛克、光纤灯、黑胡桃木等

Project Information：
Design Unit: Nanjing Luwenxing Decoration Design Studio
Design Director(Main Director): Lu Wenxing
Project Address: Anhui Hefei
Area: 2000sqm
Materials: Black and white stone, mirror mosaic, fiber optic lights, black walnut, etc.

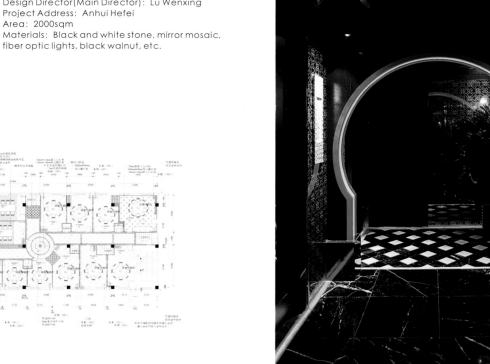

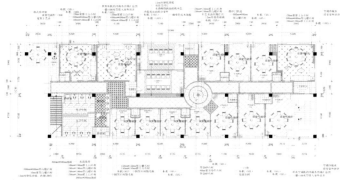

餐馆 · RESTAURANT

餐馆・RESTAURANT

餐馆 · RESTAURANT

工体享宴火锅店
GONGTI XIANGYAN POT

项目资料：
设计单位：法惟思设计
项目地址：北京工体北门右转100米
设计总监（主创）：蔡宗志
参与设计师：王为 赵志鸿 詹玉宝
摄影：孙翔宇
面积：1200m²
主要材料：石材、玻璃、钢材、木材、水泥、墙纸、铁艺

Project Information:
Design Unit: Atelier de L'AVIS
Project Address: Beijing Gongti North Gate, Turn Right 100 Miles
Design Director (Main Director): Cai Zongzhi
Involved Designer: Wang Wei, Zhao Zhihong, Zhan Yubao
Photographer: Sun Xiangyu
Area: 1200sqm
Materials: Stone, glass, steel, wood, cement, wall paper, iron

餐馆 · RESTAURANT

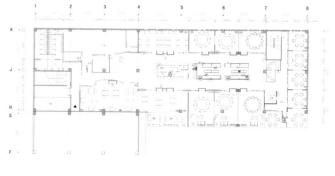

餐馆·RESTAURANT

餐馆·RESTAURANT

餐馆 · RESTAURANT

高雄新天地雅悦会馆
KAOHSIUNG NEW EARTH YAYUE CLUB

项目资料：
设计单位：暄品设计工程顾问有限公司
设计师：朱柏仰
参与设计师：何国魁
摄影：齐柏林
项目地址：高雄梦时代广场9楼
主要材料：大理石、墨镜、冲孔板、发色不锈钢
完成时间：2009年12月

Project information:
Design Unit: Xuanpin Design Engineering Consulting Co.,Ltd
Designer: Po-yang,Chu
Involed Designer: Kuo-kuei,HO
Photographer: Qi Bolin
Project Address: Gaoxiong Dream Age Square Floor Nine
Materials: Marble, Sunglasses, Punching board, Hair color of stainless steel
Completion: December 2009

餐馆 · RESTAURANT

餐馆 · RESTAURANT

黄记煌三汁焖锅中心城店

HUANGJIHUANG THREE JUICE STEW POT CENTRAL RESTAURANT

项目资料:
设计单位: 深圳市艺鼎装饰设计有限公司
设计师: 王琨
项目地址: 深圳市福田区中心城
面积: 500m²
主要材料: 大理石: 树挂冰花、古典木纹、蒙古黑、白麻石、加州金麻（桌面）
金属: 玫瑰金、黑钛金黑色木饰面、黑白马赛克、编织纹皮革、墙纸、木纹防火板

Project Information:
Design Unit: Shenzhen Yiding Decoration Design Co., Ltd.
Designer: Wang Kun
Project Address: Shenzhen Futian District Center Town
Area: 500sqm
Materials: Marble: Ice on trees, classical wood, Mongolia black, white granite, the California Gold Ma (desktop)
Metals: Rose gold, black titanium
Black wood veneer, black and white mosaics,
weaving pattern leather, wallpaper, wood fire board

鸿霖时尚餐厅
HONGLIN FASHION RESTAURANT

项目资料：
设计单位：南京浩澜设计事务所
设计总监（主创）：李浩澜
参与设计师：李娜 朱国举
摄影师：李浩澜
项目地址：南京市建邺路
项目面积：800m²
主要材料：烤漆雕花、亚麻
完成时间：2010年7月

Project Information:
Design Unit: Nanjing Haolan Design Office
Design Director(Main Director): Li Haolan
Involed Designer: Li na、Zhu Guoju
Photographer: Li Haolan
Project Address: Nanjing Jianye Road
Area: 800sqm
Materials: Paint Carved, Linen
Completion: July, 2010

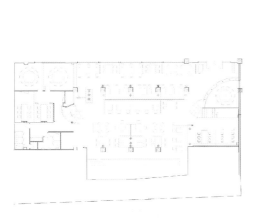

餐馆·RESTAURANT

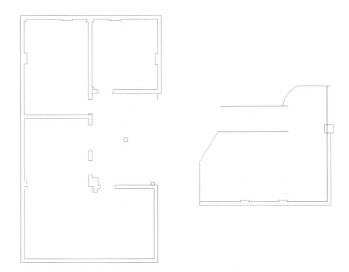

南京东部山庄 抱朴园

NANJING EASTERN VILLA BAOPU PARK

项目资料：
设计单位：南京正午阳光装饰设计工作室
设计师：张有东
摄影：张有东
项目地址：南京栖霞区312国道旁
面积：700m²
主要材料：皮纹砖、仿古砖、不锈钢、灰镜、水曲柳染色、砂岩佛雕、小青瓦、峒石、东阳木雕、金属画

Project Information:
Design Unit: Nanjing Midday Sunny Decoration Design Studio
Designer: Zhang Youdong
Photographer: Zhang Youdong
Project Address: Nanjing Qixia District beside 312 State Road
Area: 700sqm
Materials: Striae brick, antique brick, stainless steel, gray mirror, stained ash, sandstone Buddhist sculpture, small gray tiles, Dong stone, Dongyang wood carving, metal Painting

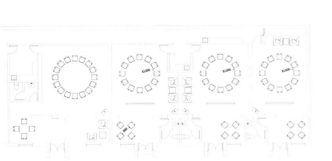

餐馆 · RESTAURANT

闽都别景 · 燕鲍鱼翅会所

MINDU BIEJING.ABALONE AND FIN CLUB

项目资料:
设计单位：福建国广一叶建筑装饰设计工程有限公司
设计师：杨思聪 陈茂春 金舒扬 刘国铭
项目地址：福州大饭店顺风楼三层
面积：1300m²

Project Information:
Design Unit: Fujian Guoguang Yiye Architectural Decoration Design Engineering Co., Ltd.
Designer: Yang Sicong, Chen Maochun, Jin Shuyang, Liu Guoming
Project Address: Fuzhou Hotel Shunfeng Building Floor Three
Area: 1300sqm

餐馆·RESTAURANT

餐馆 · RESTAURANT

餐馆・RESTAURANT

餐馆 · RESTAURANT

望湘园金桥店
XIANG GARDEN JINQIAO BRANCH

项目资料：
设计单位：上海沈敏良室内设计有限公司
设计师：沈敏良
摄影师：蔡峰
项目地址：上海市浦东金桥
面积：674m²
主要材料：木纹砖、镜面不锈钢、黄镜、灰镜、壁纸

Project Information:
Design Unit: Shanghai Shenminliang Interior Design Co., Ltd.
Designer: Shen Minliang
Photographer: Caifeng
Project Address: Shanghai Pudongjin Bridge
Area: 674sqm
Materials: Wood tiles, stainless steel mirror, yellow mirror, gray mirror, wallpaper

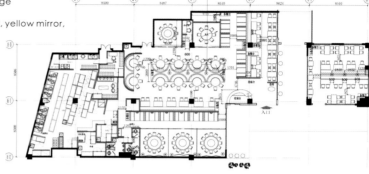

餐馆 · RESTAURANT

星愿时尚餐厅
STAR WISH FASHION RESTAURANT

项目资料：
设计单位：南京浩澜设计事务所
设计总监（主创）：李浩澜
参与设计师：李娜 朱国举
摄影师：李浩澜
项目地址：南京王府大街
面积：300m²
主要材料：灰镜、马赛克
完成时间：2010年5月

Project Information:
Design Unit: Nanjing Haolan Design Office
Design Director(Main Director): Li Haolan
Involed Designer: Li na、Zhu Guoju
Photographer: Li Haolan
Project Address: Nanjing Wangfu Street
Area: 300sqm
Materials: Gray Mirror, Mosaics
Completion: May, 2010

餐馆 · RESTAURANT

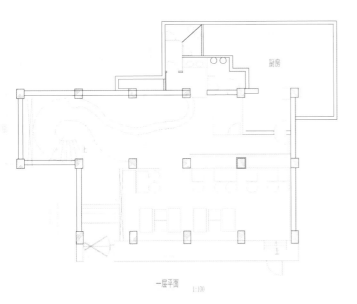

一层平面 1:100

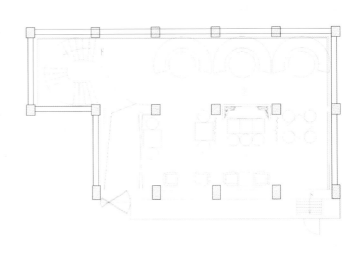

二层平面 1:100

餐馆 · RESTAURANT

餐馆·RESTAURANT

扬州牛排城
（四望亭店）
YANGZHOU STEAK TOWN
(SIWANG PAVILION BRANCH)

项目资料：
设计单位：上瑞元筑设计制作有限公司
设计师：孙黎明 胡红波
摄影师：戴俊峰
项目地址：四望亭路28号
面积：800m²
主要材料：仿古迭石、地砖、中花白大理石、橡木板、迭纹墙纸
完成时间：2009年12月

Project Information:
Design Unit: Shangrui Yuanzhu Design Making Co., Ltd.
Designer: Sun Liming, Hu Hongbo
Photographer: Dai Junfeng
Address: Siwangting Road No.28
Area: 800sqm
Materials: antique stone, floor tiles, the gray marble, oak board, wallpaper
Completion: December, 2009

餐馆 · RESTAURANT

餐馆 · RESTAURANT

尚品世家
SHANGPIN FAMILY

项目资料：
设计单位： 上瑞元筑设计制作有限公司
设计师： 冯嘉云 高毅南
摄影师： 戴俊峰
项目地址： 无锡市湖滨路酒吧街
面积： 600m²
主要材料： 风化木材、老红砖、墙纸、青砖、水曲柳
完成时间： 2010年02月

Project Information:
Design Unit: Shangrui Yuanzhu Design Making Co., Ltd.
Designer: Feng Jiayun, Gao Yinan
Photographer: Dai Junfeng
Address: Wuxi Hubin Road Bar Street
Area: 600 sqm
Materials: Weathered wood, old red brick, wallpaper, brick, Shuiquliu
Completion: February, 2010

餐馆·RESTAURANT

WGV CAFETERIA

项目资料：
设计单位：Ippolito Fleitz Group GmbH Identity Architects
设计师：Peter Ippolito, Gunter Fleitz, Mathias Modinger, Lena Noh, Tim Lessmann
摄影师：Zooey Braun
客户：Württembergische Gemeinde
项目地址：Hauptstatter Straße 100, 70174 Stuttgart

Project Information:
Design Unit: Ippolito Fleitz Group GmbH Identity Architects
Designer: Peter Ippolito, Gunter Fleitz, Mathias Modinger, Lena Noh, Tim Lessmann
Photographer: Zooey Braun
Client: Württembergische Gemeinde
Address: Hauptstatter Straße 100, 70174 Stuttgart

餐馆 · RESTAURANT

美丽华酒店员工餐厅
THE MIRA STAFF RESTAURANT

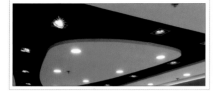

项目资料：
设计单位：DPWT Design Ltd.
设计团队：陈轩明 白雪梅
客户：美丽华集团
项目地址：香港九龙尖沙咀弥敦道
面积：355m²
设计风格：现代、有活力、启发灵感
主要材料：清镜、不锈钢、拼花木地板、重点强调的暖色调子喷漆墙身与塑料薄膜照明
完成时间：2009年5月

Project Information:
Design Unit: DPWT Design Ltd.
Designer: Arthur Chan, Jenny Bai
Client: The Miramar Group
Project Address: 118-130 Nathan Road, Tsim Sha Tsui, Hong Kong
Area: 355sqm.
Design Style: Modern, Energetic, and Inspiring
Materials: Mirror, stainless steel, wooden parquet flooring, vibrant, warm tone colours' spray paint wall for accent highlights.
Completion: May 2009

餐馆 · RESTAURANT

新紫阳大酒楼
NEW ZIYANG RESTAURANT

项目资料：
设计单位：福建国广一叶建筑装饰设计工程有限公司
设计师：金舒扬 林祥通 林圳钦 刘国铭 陈垚 陈剑英
项目地址：福州福新路
面积：3500m²
主要材料：木饰面、原木、软包、镜面、不绣钢、金箔等

Project Information:
Design Unit: Fujian Guoguang Yiye Architectural Decoration Design Engineering Co., Ltd.
Designer: Jin Shuyang, Lin Xiangtong, Lin Zhenqin, Liu Guoming, Chen Yao, Chen Jianying
Project Address: Fuzhou Fuxin Road
Area: 3500 sqm
Materials: Wood veneer, wood, soft bag, mirror, stainless steel, gold, etc.

餐馆 · RESTAURANT

餐馆 · RESTAURANT

餐馆·RESTAURANT

隐泉HATSUNE 日式料理

SECRET SPRING HATSUNE
JAPANESE CUISINE

项目资料：
设计单位：法惟思设计
项目地址：三里屯 Village
设计总监（主创）：蔡宗志
参与设计师：赵志鸿 詹玉宝
摄影师：孙翔宇
主要材料：毛石、老木、不锈钢、玻璃、木条、水泥、瓷器、大卵石、锈铁、有机玻璃、墙纸、布艺

Project Information:
Design Unit: Atelier de L'AVIS
Project Address: Sanlitun
Designer: Cai Zongzhi
Involved Designer: Zhao Zhihong, Zhan Yubao
Photographer: Sun Xiangyu
Materials: Rubble, old wood, stainless steel, glass, wood, cement, porcelain, large pebbles, rusty iron, glass, wallpaper, fabric

餐馆·RESTAURANT

01 入口
02 寿司台
03 接待台
04 **VIP**
05 餐区
06 酒吧
07 厨房

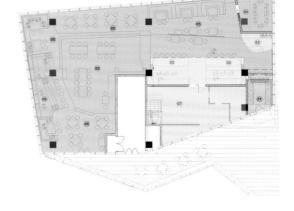

餐馆·RESTAURANT

餐馆 · RESTAURANT

扬州牛排城（珍园店）

YANGZHOU STEAK TOWN (ZHENYUAN BRANCH)

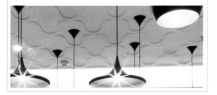

项目资料：
设计单位：上瑞元筑设计制作有限公司
设计师：孙黎明 胡红波
摄影师：戴俊峰
项目地址：邗江区京华城路168号
面积：700m²
主要材料：橡木板、地砖、中花白大理石、铜镜、黑色钢琴漆、松木板、车边茶镜、白镜
完成时间：2009年05月

Project Information:
Design Unit: Shangrui Yuanzhu Design Making Co., Ltd.
Designer: Sun Liming
Photographer: Dai Junfeng
Address: Hanjiang District Jinghuacheng Road No. 168
Area: 700sqm
Materials: Oak panels, floor tiles, the gray marble, bronze mirrors, black piano lacquer, pine boards, car side tinned mirror, white mirror
Completion: May 2009

餐馆 · RESTAURANT

餐馆·RESTAURANT

天度会
TIANDU CLUB

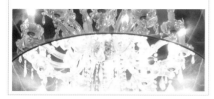

项目资料：
设计单位：福州林开新室内设计有限公司

Project Information:
Design Unit: Fuzhou Linkaixin Interior Design Co., Ltd.

餐馆・RESTAURANT

餐馆 · RESTAURANT

台米台菜（五四店）
TAIMI TAICAI (FIVE FOUR BRANCH)

项目资料：
设计单位：福建国广一叶建筑装饰设计工程有限公司
设计师：金舒扬 刘国铭 林圳钦
项目地址：福州
面积：约900m²
主要材料：文化石、原木、玻璃
方案审定：叶斌

Project Information:
Design Unit: Fujian Guoguang Yiye Architectural Decoration Design Engineering Co., Ltd.
Designer: Jin Shuyang, Liu Guoming, Lin Zhenqin
Project Address: Fuzhou
Area: 900sqm
Materials: Culture stone, wood, glass
Validation program: Ye Bin

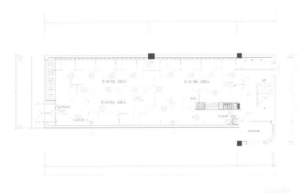

餐馆·RESTAURANT

食为天菜馆
FOOD IS GOD RESTAURANT

项目资料：
设计单位：上海五铭建筑装饰设计有限公司
设计师：顾宇光
摄影师：顾宇光
项目地址：中国江苏射阳
面积：1200m²
主要材料：钢管、木材、银镜
完成时间：2009年9月

Project Information:
Design Unit: Shanghai Wuming Architectural Decoration Design Co., Ltd.
Designer: Gu Yuguang
Photographer: Gu Yuguang
Project Address: China Jiangsu Sheyang
Area: 1200sqm
Materials: Steel, Wood, Silver Mirror
Completion: September 2009

餐馆 · RESTAURANT

食尚廊桥火锅
FOOD BRIDGES POT

项目资料：
设计单位：成都全兴广告装饰有限公司
设计总监：肖波

Project Information:
Design Unit: Chengdu Quanxing Advertisement Decoration Co., Ltd.
Design Director: Xiao Bo

餐馆 · RESTAURANT

餐馆·RESTAURANT

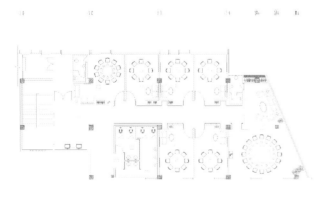

餐馆 · RESTAURANT

LUCK 料理店
LUCK RESTAURANT

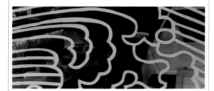

项目资料：
设计单位：大连工业大学
主设计师：张健
参与设计师：郭佳 李禹
主要材料：金属马赛克、石英石、文化石、镜面玻璃、TOTO洁具、三雄照明灯具
面积：100m²
完成时间：2010年7月

Project Information:
Design Unit: Dalian Industry University
Main Designer: Zhang Jian
Involed Designer: Guo Jia, Li Yu
Materials: Metallic mosaic, quartz stone, culture stone, mirror glass, TOTO sanitary ware, Sanxiong lighting
Area: 100sqm
Completion: July 2010

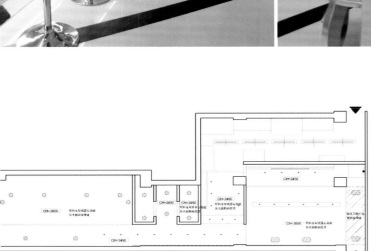

天花布置图　　　　　　　　平面布置图

餐馆 · RESTAURANT

连云港腾轩火锅店
LIANYUN HARBOUR TENGXUAN POT SHOP

项目资料：
设计单位：北京丽贝亚建筑装饰工程有限公司
设计总监（主创）：金哲秀
参与设计师：刘菊
摄影师：徐盟
项目地址：江苏连云港
主要材料：有色涂料、有色钢琴漆、茶镜、冰川纹矿棉板、水泥、浇筑、硬包布料、壁纸、茶玻、黑玻、艺术玻璃、霸王花石材和地砖、木地板

Project Information:
Design Unit: Beijing Libeiya Architectural Decoration Engineering Co., Ltd.
Designer: Jin Zhexiu
Involved Designer: Liu Ju
Photographer: Xu Meng
Project Address: Jiangsu Lianyun Harbor
Materials: Colored paint, colored piano-paint, tea mirror, ice grain mineral wool board, cement, casting, hard pack fabrics, wallpaper, tea glass, black glass, art glass, Bawanghua stone and flower, wooden floor

餐馆 · RESTAURANT

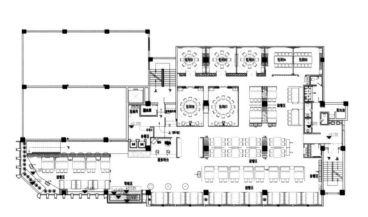

餐馆·RESTAURANT

净雅阳光海岸 嘉和店
JINGYA SUNLIGHT COAST JIAHE SHOP

项目资料：
设计单位：北京丽贝亚建筑装饰工程有限公司
设计总监（主创）：迟凯
参与设计师：杜春亮
摄影师：徐盟
项目地址：北京市
主要材料：石材、玻璃、木饰面

Project Information:
Design Unit: Beijing Libeiya Architectural Decoration Engineering Co., Ltd.
Designer: Chi Kai
Involved Designer: Du Chunliang
Photographer: Xu Meng
Project Address: Beijing
Materials: Stone, Glass, Wood Finishes

餐馆 · RESTAURANT

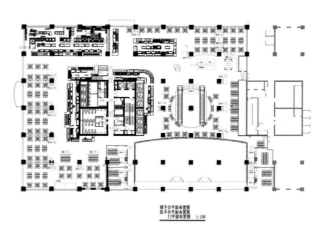

餐馆 · RESTAURANT

杭州名爵咖啡馆
HANGZHOU MINGJUE COFFEE SHOP

项目资料：
设计单位：杭州屹展室内设计工作室
设计总监（主创）：蒋捍明 肖懿展
参与设计师：黄道永
摄影：訾向平
项目地址：杭州
面积：890m²
主要材料：水曲柳饰面板、大理石、仿古砖
完成时间：2010年10月

Project Information:
Design Unit: Hangzhou Yizhan Interior Design Studio
Design Director: Jiang Hanming, Xiao Yizhan
Involved Designer: Huang Daoyong
Photographer: Zi Xiangping
Project Address: Hangzhou
Area: 890sqm
Materials: Shuiquliu veneer, marble, antique tiles
Completion: October 2010

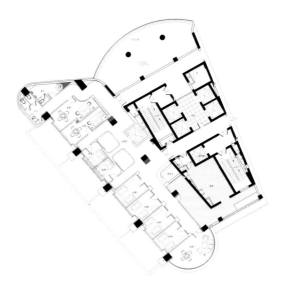

餐馆 · RESTAURANT

汉府餐厅
HAN RESTAURANT

项目资料：
设计单位：善水堂创意设计机构
设计师：朱伟
摄影师：王一威
主要材料：竹板、老杉木、红色条砖、草编墙纸

Project Information:
Design Unit: Shanshuitang Creative Design Institution
Designer: Zhu Wei
Photographer: Wang Yiwei
Materials: Bamboo, old fir, red bar brick, straw wallpaper

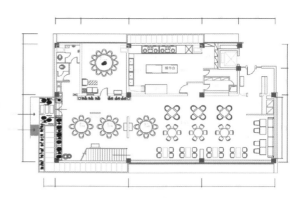

餐馆 · RESTAURANT

干锅轩
GRIDDLE RESTAURANT

项目资料：
设计单位：深圳华空间机构

Project Information:
Design Unit: Shenzhen Hua Space Constitution

餐馆 · RESTAURANT

餐馆 · RESTAURANT

夜朗蛙工程
NIGHT FROG ENGINEERING

项目资料：
设计师：刘秀梅 陈旭东
摄影师：陈旭东

Project Information:
Designer: Liu Xiumei Chen Xudong
Photographer: Chen Xudong

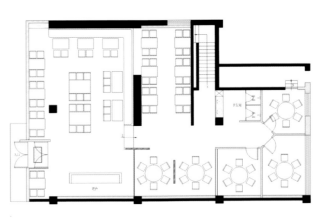

餐馆 · RESTAURANT

餐馆·RESTAURANT

会仙莜面馆
HUIXIANYOU NOODLES

项目资料：
设计单位：北京屹峰旭辰空间设计有限公司
设计师：韩金锁
项目地址：山西省孝义市
面积：600m²
完成时间：2010年1月

Project information:
Design Unit: Beijing Yifeng Xuchen Space Design Co., Ltd
Designer: Han Jinsuo
Project Address: Shanxu Xiaoyi
Area: 600sqm
Completion: January 2010

THE NIGHT MARKET

THE NIGHT MARKET

项目资料：
设计单位：Michael Young
设计师：Alexi Robinson Interiors and Michael Young

Project Information:
Design Unit: Michael Young
Designer: Alexi Robinson Interiors and Michael Young

餐馆 · RESTAURANT

SECOND HOME KITCHEN AND BAR

SECOND HOME KITCHEN AND BAR

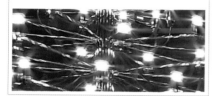

项目资料：
设计单位： Andre Kikoski Architect
摄影师： Eric Laignel
面积： 5000平方英尺
项目地址： 150 Clayton Lane Denver, Co80206
完成时间： 2008年3月

Project Information:
Design Unit: Andre Kikoski Architect
Photographer: Eric Laignel
Area: 5000SF
Address: 150 Clayton Lane Denver, Co80206
Completion: March, 2008

餐馆·RESTAURANT

APOSTROPHE

APOSTROPHE

项目资料：
设计单位：SHH Architects + Interiors + Design Consultants
设计师：SHH
设计团队：Neil Hogan, Brendan Heath, Adam Woodward
摄影师：Francesca Yorke
客户：Apostrophe

Project Information:
Design Unit: SHH Architects + Interiors + Design Consultants
Designer: SHH
Design Team: Neil Hogan, Brendan Heath, Adam Woodward
Photographer: Francesca Yorke
Client: Apostrophe

餐馆 · RESTAURANT

大渔铁板烧喜荟城店

DAYU IRON XIHUI RESTAURANT

项目资料：
设计单位： 深圳市艺鼎装饰设计有限公司
设计师： 王锟
项目地址： 深圳罗湖区喜荟城
面积： 400m²
主要材料： 黑木纹、蒙古黑、树挂冰花大理石、咖啡影木饰面、玫瑰金、原木晕画

Project Information:
Design Unit: Shenzhen Yiding Decoration Design Co., Ltd.
Designer: Wang Kun
Project Address: Shenzhen Luohu District Xihui Town
Area: 400sqm
Materials: Black wood, Mongolia Black, Tree Hanging Ice Marble, Coffee Shadow Wood Veneer, Rose Gold, Wood Painted Halo

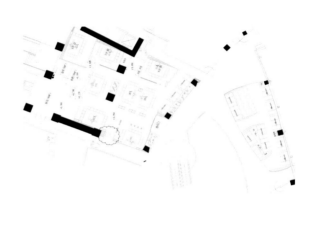

餐馆·RESTAURANT

北京湘君府酒楼
BEIJING XIANGJUN RESTAURANT

项目资料：
设计单位：香港睿·设计工程有限公司
设计师：吴睿 顾建平 刘晓彤
参与设计团队：睿·设计事务所
项目地址：北京市广安门桥高新大厦
建筑面积：4700m²

Project Information：
Design Unit: Hong Kong Rui. Design Engineering Co., Ltd.
Designer: Wu Rui, Gu Jianping, Liu Xiaotong
Involed Design Team: Rui.Design Engineering Co., Ltd.
Project Address: Beijing Guang'anmen Bridge Gaoxin Building
Area: 4700sqm

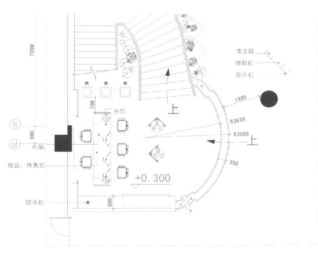

餐馆 · RESTAURANT

餐馆 · RESTAURANT

餐馆 · RESTAURANT

餐馆 · RESTAURANT

YELLOW DELI
YELLOW DELI

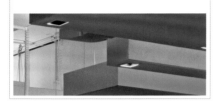

项目资料：
设计单位：SÖHNE & PARTNER Architect

Project Information:
Design Unit: SÖHNE & PARTNER Architects

餐馆·RESTAURANT

JAMIE'S ITALIAN

JAMIE'S ITALIAN

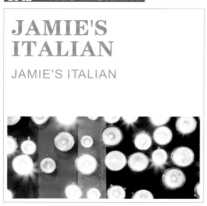

项目资料：
设计单位：Blacksheep
摄影师：Gareth Gardner

Project Information:
Design Unit: Blacksheep
Photographer: Gareth Gardner

餐馆 · RESTAURANT

餐馆 · RESTAURANT

巴比肯 FOOD-HALL 和酒廊
BARBICAN FOODHALL & LOUNGE

项目资料：
设计单位：SHH Architects & Design Censaltants
设计师：SHH
摄影师：Gareth Gardner
Caroline Collett

Project Information:
Design Unit: SHH Architects & Design Censaltants
Designer: SHH
Photography: Gareth Gardner
Caroline Collett

餐馆・RESTAURANT

餐馆 · RESTAURANT

NATURE CAFÉ AND NATURE BOUTIQUE

NATURE CAFÉ AND NATURE BOUTIQUE

项目资料:
设计单位: Reich+Petch Design International
艺术总监: Edmund Li
主要设计师: Tony Reich
平面设计师: Vivien Chow
室内设计师: Cathy Misiaszek
资深技师: Mark Catchpole
设计师: Marc Kyriakou
摄影师: Martin Lipman
面积: 232m²

Project Information:
Design Unit: Reich+Petch Design International
Art Director: Edmund Li
Principal: Tony Reich
Graphic Designer: Vivien Chow
Interior Designer: Cathy Misiaszek
Senior Technologist: Mark Catchpole
Designer: Marc Kyriakou
Photographer: Martin Lipman
Area: 232sqm

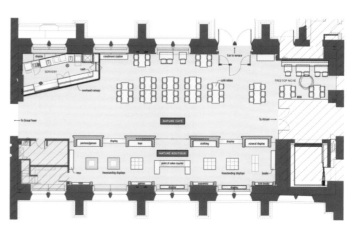

餐馆 · RESTAURANT

餐馆 · RESTAURANT

SUPPERCLUB SINGAPORE

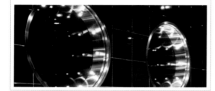

项目资料:
设计单位: Concrete Architectural Associates
项目团队: Rob Wagemans, Erik van Dillen, Jan Paul Scholtmeijer
摄影师: Frank Pinckers
客户: IQ creative
项目地址: Odeon Towers, 331 North Bridge Road, Singapore
面积: 1300m²

Project Information:
Design Unit: Concrete Architectural Associates
Project Team: Rob Wagemans, Erik van Dillen, Jan Paul Scholtmeijer
Photographer: Frank Pinckers
Client: IQ creative
Project Address: Odeon Towers, 331 North Bridge Road, Singapore
Area: 1300sqm

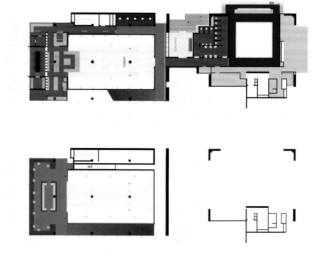

餐馆 · RESTAURANT

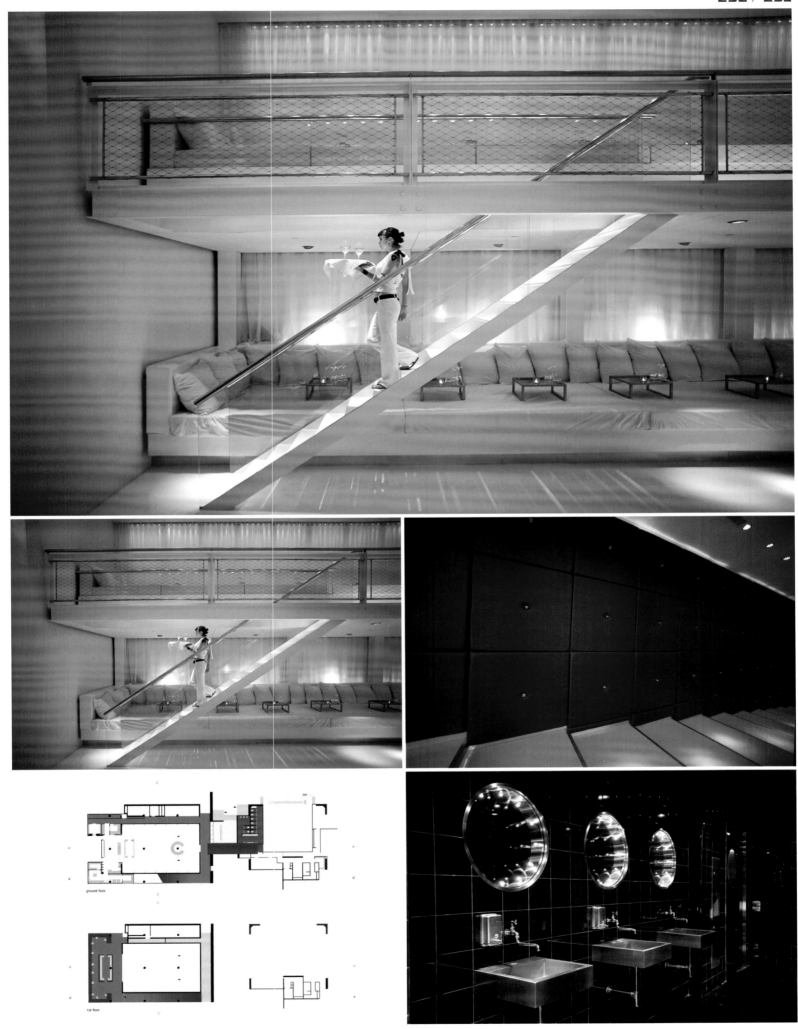

餐馆 · RESTAURANT

MOC MOC

项目资料：
设计单位：GRO Architects
创意总监：Rchard Garber + Nicole Robertson, Justin Foster
摄影师：Fabian Birgfeld

Project Information:
Design Unit: GRO Architects
Creative Director: Rchard Garber + Nicole Robertson, Justin Foster
Photographer: Fabian Birgfeld

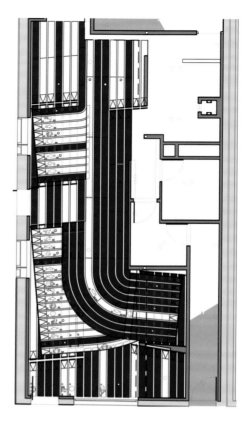

餐馆 · RESTAURANT

ESPRESSO IN EXISTING SPACE IJBURG

ESPRESSO IN EXISTING SPACE IJBURG

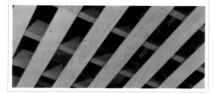

项目资料：
设计单位：Studio Ramin Visch
摄影师：Jeroen Musch
项目团队：Ramin Visch, Femke Poppinga
客户：Mr. Rick W. Woertman
面积：70 m²

Project Information:
Design Unit: Studio Ramin Visch
Photography: Jeroen Musch
Project team: Ramin Visch, Femke Poppinga
Client: Mr. Rick W. Woertman
Area: 70sqm

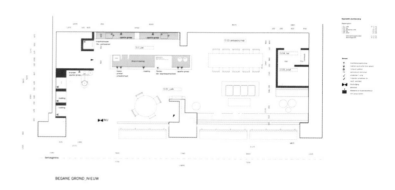

餐馆・RESTAURANT

福师傅连锁餐厅
FU MASTER CHAIN RESTAURANTS

项目资料：
设计单位：深圳华空间机构

Project Information:
Design Unit: Shenzhen Hua Space Constitution

餐馆 · RESTAURANT

餐馆 · RESTAURANT

唐山凤凰园烤鸭店 唐丰路店

TANGSHAN PHOENIX ROASTED DUCK RESTAURANT-FENGLU BRANCH

项目资料：
设计单位：香港睿·设计工程有限公司
设计师：吴睿 顾建平 刘晓彤
参与设计团队：睿·设计工程有限公司
摄影师：司徒颖
项目地址：唐山市唐丰路
面积：5800m²
主要材料：石材、夹绢玻璃、青石雕刻、粉镜、仿皮面硬包、木做、黑碳钢、幻彩金箔、壁纸、仿古磁砖、地毯

Project Information:
Design Unit: Hong Kong Rui. Design Engineering Co., Ltd.
Designer: Wu Rui, Gu Jianping, Liu Xiaotong
Involed Design Team: Rui.Design Engineering Co., Ltd.
Photographer: Situ Ying
Project Address: Tangshan Tangfeng Road
Area: 5800sqm
Materials: Stone, silk glass, bluestone carving, powder mirror, leather face hard pack, wood made, black carbon, Symphony gold, Wallpaper, antique tiles, carpet

餐馆 · RESTAURANT

餐馆 · RESTAURANT

餐馆 · RESTAURANT

餐馆 · RESTAURANT

南京东佳会馆宴会厅

NANJING DONGJIA HALL BALLROOM

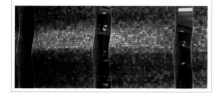

项目资料：
设计单位：南京正午阳光装饰设计工作室
设计师：张有东
摄影：张有东
项目地址：南京六合开发区
主要材料：金箔墙纸、不锈钢、灰镜、清玻磨花、烤漆玻璃透光

Project Information:
Design Unit: Nanjing Midday Sunny Decoration Design Studio
Designer: Zhang Youdong
Photographer: Zhang Youdong
Project Address: Nanjing Liuhe Develop District
Materials: Golden wallpaper, stainless steel, gray mirror, clear glass wear flowers, translucent glass paint

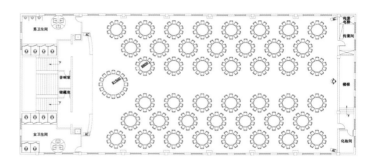

二层平面布置图

餐馆 · RESTAURANT

KOSOMO咖啡店
KOSOMO COFFEE SHOP

项目资料：
设计单位：深圳华空间机构
设计师：熊华阳
项目地址：东莞

Project Information:
Design Unit: Shenzhen Hua Space Constitution
Designer: Xiong Huayang
Project Address: Dongguan

餐馆·RESTAURANT

杭州泰晤士咖啡馆
HANGZHOU THAMES COFFEE SHOP

项目资料：
设计单位：杭州屹展室内设计工作室
设计总监（主创）：蒋捍明、肖懿展
参与设计师：胡艳杰
摄影师：訾向平
项目地址：杭州滨江区
面积：1100m²
主要材料：深啡网纹大理石、水曲柳
完成时间：2009年8月

Project Information：
Design Unit: Hangzhou Yizhan Interior Design Studio
Design Director(Main Director): Jiang Hanming, Xiao Yizhan
Involved Designer: Hu Yanjie
Photographer: Zi Xiangping
Project Address: Hangzhou Binjiang District
Area: 1100sqm
Materials: Reticulate dark brown marble, Shuiquliu
Completion: August 2009

餐馆·RESTAURANT

湘乐汇莘庄仲盛店
XIAONG XINZHUANG BRANCH

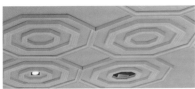

项目资料：
设计单位：上海沈敏良室内设计有限公司
设计总监（主创）：沈敏良
参与设计团队：上海沈敏良室内设计有限公司
摄影师：蔡峰
项目地址：上海市闵行区莘庄仲盛广场
面积：742m²
主要材料：茶镜、黄镜、黑镜、镜面不锈钢、墙纸、黄洞石、陶瓷马赛克

Project Information:
Design Unit: Shanghai Shenminliang Interior Design Co., Ltd.
Designer: Shen Minliang
Involved Team: Shanghai Shenminliang Interior Design Co., Ltd.
Photographer: Caifeng
Project Address: Shanghai Minxing District Zizhuang Zhongsheng Sqaure
Area: 742sqm
Materials: Tea mirror, yellow mirror, black mirror, mirror stainless steel, wallpaper, yellow cave-stone, ceramic mosaic

餐馆·RESTAURANT

餐馆 · RESTAURANT

SALON DES SALUTS

SALON DES SALUTS

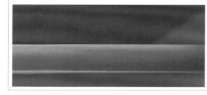

项目资料：
设计单位：sinato
设计师：Chikara Ohno / sinato
摄影师：Toshiyuki Yano
客户：NATURAL EARTH Inc.
项目地址：1-4-20-1F Nishiazabu Minato-ku Tokyo 106-0031 Japan
面积：53.51m²
完成时间：2009年10月

Project Information:
Design Unit: sinato
Designer: Chikara Ohno / sinato
Photographer: Toshiyuki Yano
Client: NATURAL EARTH Inc.
Address: 1-4-20-1F Nishiazabu Minato-ku Tokyo 106-0031 Japan
Area: 53.51sqm
Completion: October 2009

餐馆 · RESTAURANT

NEVY
NEVY

项目资料：
设计单位：Concrete Architectural Associates
项目团队：Rob Wagemans, Janpaul scholtmeijer, Jari van lieshout, Erik van Dillen
摄影师：ewout huibers
客户：IQ creative
项目地址：westerdoksdijk 40, 1013 Ae, Amsterdam
面积：220m²

Project Information:
Design Unit: Concrete Architectural Associates
Project Team: Rob Wagemans, Janpaul scholtmeijer, Jari van lieshout, Erik van Dillen
Photographe: ewout huibers
Client: Q Creative
Address: westerdoksdijk 40, 1013 AE, Amsterdam
Area: 220sqm

P*ONG DESSE – RT BAR
P*ONG DESSERT BAR

项目资料:
设计单位: Andre Kikoski Architect
设计师: Andre Kikoski, AIA, LEED AP
摄影师: Eric Laignel
项目地址: New York

Project Information:
Design Unit: Andre Kikoski Architect
Designer: Andre Kikoski, AIA, LEED AP
Photographer: Eric Laignel
Project Address: New York

餐馆 · RESTAURANT

BRASSERIE WITTEVEEN

BRASSERIE WITTEVEEN

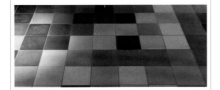

项目资料：
设计单位：Concrete Architectural Associates
项目团队：Rob Wagemans，Ulrike Lehner，Joyce Kelder，Erik van Dillen
平面设计：Sofie Ruytenberg
摄影师：Ewout Huibers
项目地址：Ceintuurbaan 256-260, 1072 GH Amsterdam
面积：400m²

Project Information:
Design Unit: Concrete Architectural Associates
Project Team: Rob Wagemans, Ulrike Lehner, Joyce Kelder, Erik van Dillen
Graphic Designer: Sofie Ruytenberg
Photographer: Ewout Huibers
Address: Ceintuurbaan 256-260, 1072 GH Amsterdam
Area: 400sqm

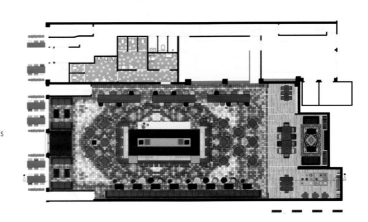

餐馆·RESTAURANT

WALL & WATER RESTAURANT

项目资料：
设计单位：Rockwell Group
摄影师：Courtesy of Andaz Wall Street
客户：Hyatt Hotels Corporation, The Hakimian Organization

Project Information:
Design Unit: Rockwell Group
Photographer: Courtesy of Andaz Wall Street
Client: Hyatt Hotels Corporation, The Hakimian Organization

餐馆·RESTAURANT

RESTAURANT TUSEW SWEDEN RAMUNDBERSER

RESTAURANT TUSEW SWEDEN RAMUNDBERSER

项目资料：
设计单位：Murman Arkitekter AB
建筑团队：Hans Murman Ulla Alberts
摄影师：Aue-e-sm Undman/ Hans Murman

Project Information:
Design Unit: Murman Arkitekter AB
Architect Team: Hans Murman Ulla Alberts
Photographer: Aue-e-sm Undman/ Hans Murman

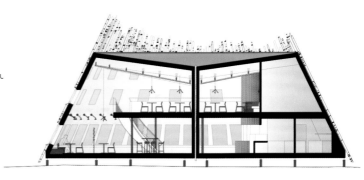

SEKTION

餐馆 · RESTAURANT

FORNERIA SAN PAOLO

FORNERIA SAN PAOLO

项目资料：
设计单位：studio mk27
设计师：marcio kogan
参与设计师：lair reis + diana radomysler + studio mk27 team
摄影师：Romulo fialdini
主要材料：大理石马塞克、木质天花板
面积：320m²

Project Information：
Design Unit: studio mk27
Designer: marcio kogan
Involed Designer: lair reis + diana radomysler + studio mk27 team
Photographer: Romulo fialdini
Materials: marble mosaic, wooden ceiling
Area: 320sqm

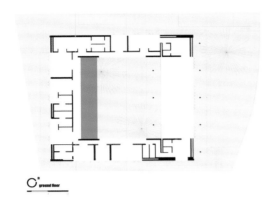

CAFÉ3
CAFÉ 3

项目资料:
设计单位: Andre Kikoski Architect
摄影师: Peter Aaron at ESTO, Eric Laignel

Project Information:
Design Unit: Andre Kikoski Architect
Photographer: Peter Aaron at ESTO, Eric Laignel

ENLARGED FLOOR PLAN

餐馆 · RESTAURANT

CAFFÈ DEI MUSEI

CAFFÈ DEI MUSEI

项目资料：
设计单位：Andrea Meirana Architects
创意总监：arch. Andrea Meirana
项目团队：arch. Luca Parodi, arch. Ilaria Cargiolli, designer Mauro Valsecchi, arch. Magalie Ehret.
摄影师：Alberto Ferrero

Project Information:
Design Unit: Andrea Meirana Architects
Creative Director: arch. Andrea Meirana
Project Team: arch. Luca Parodi, arch. Ilaria Cargiolli, designer Mauro Valsecchi, arch. Magalie Ehret.
Photographer: Alberto Ferrero

LEGENDA
01 SLIDING GLASS ENTRANCE
02 COFFEE BAR
03 COUNTER
04 CASH DESK
05 BREAKFAST/BUFFET TABLES
06 WINE EXHIBITOR
07 PRIVATE WINE CELLAR
08 SCENOGRAPHIC WALL
09 BATHROOM
10 KITCHEN

餐馆 · RESTAURANT

ADOUR ALAIN DUCASSE AT THE ST. REGIS

ADOUR ALAIN DUCASSE AT THE ST. REGIS

项目资料：
设计单位：Rockwell Group
摄影师：Bruce Buck
客户：Alain Ducasse

Project Information:
Design Unit: Rockwell Group
Photographer: Bruce Buck
Client: Alain Ducasse

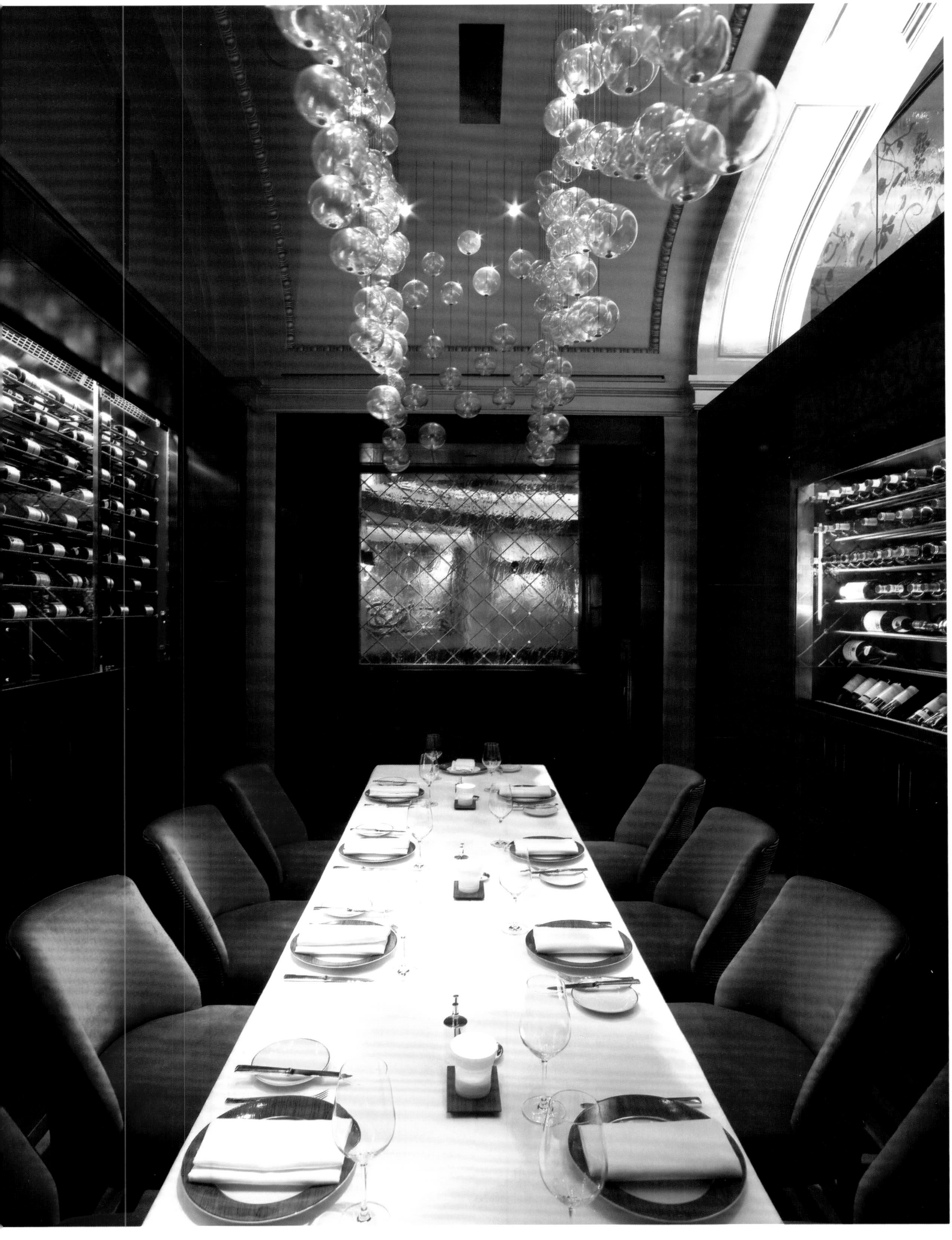

餐馆·RESTAURANT

海得利餐饮酒店
SEA DELI DINING HOTEL

项目资料：
设计单位：上海同济创意坊李景光空间
设计工作室

Project information:
Design Unit: Shanghai Tongji Creative Lijingguang Space Design Studio

餐馆 · RESTAURANT